GARDEN PLANNER
and
SEED INVENTORY JOURNAL

Created by:

Beth L. Wickstrum

HomeTrees Press
Ojai · California

© Copyright 2023

This Garden Planner Belongs to:

_____

Year: _____

Life begins the day you start a garden.

—Chinese Proverb

How to use this Planner

This Garden Planner is a culmination of informational items we found that we needed to track our garden tasks. Use the logs that are pertinent to you and your specific gardening needs.

We have included a section to keep your ever-growing seed inventory organized. There is also a Garden Layout page where you can draw a layout of your garden and label your beds. Helpful pages include: a Seed Germination Log, a page for each garden bed with specific information on the crops that you planted, which companion plants you used, planting dates, germination rates, and crop rotations.

The Seed Inventory pages will help you track your seed supply and whether you need to reorder specific seeds. The Seed Germination Log was designed to track how you initially started your seeds, whether indoors or directly in the soil, the date they were planted and their designated garden bed.

We have provided a Supplier Contact List to keep all of your important resources in one place and a section for Garden Notes where you can jot down pertinent things such as pests and diseases that affect your plants.

Happy Gardening!

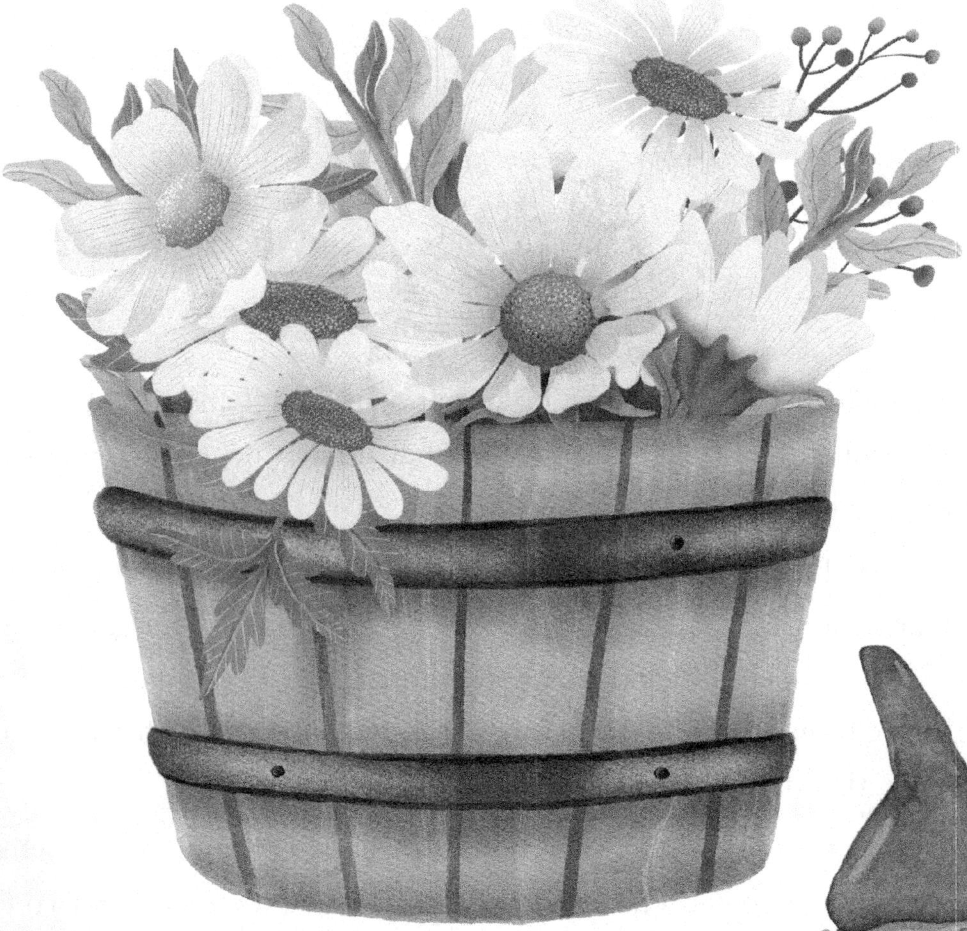

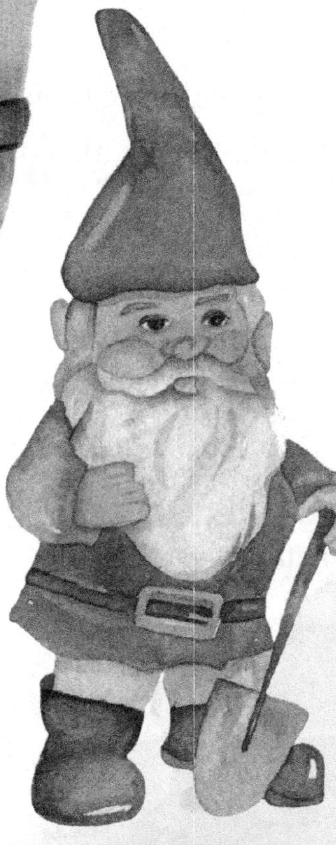

Seed Inventory

SEED INFORMATION

Seed Name:

Year on Packet:

Year Purchased:

Company:

Approx. Seeds Left:

Reorder: ○ YES ○ NO

SEED INFORMATION

Seed Name:

Year on Packet:

Year Purchased:

Company:

Approx. Seeds Left:

Reorder: ○ YES ○ NO

SEED INFORMATION

Seed Name:

Year on Packet:

Year Purchased:

Company:

Approx. Seeds Left:

Reorder: ○ YES ○ NO

SEED INFORMATION

Seed Name:

Year on Packet:

Year Purchased:

Company:

Approx. Seeds Left:

Reorder: ○ YES ○ NO

SEED INFORMATION

Seed Name:

Year on Packet:

Year Purchased:

Company:

Approx. Seeds Left:

Reorder: ○ YES ○ NO

SEED INFORMATION

Seed Name:

Year on Packet:

Year Purchased:

Company:

Approx. Seeds Left:

Reorder: ○ YES ○ NO

Seed Inventory

SEED INFORMATION

Seed Name:

Year on Packet:

Year Purchased:

Company:

Approx. Seeds Left:

Reorder: ○ YES ○ NO

SEED INFORMATION

Seed Name:

Year on Packet:

Year Purchased:

Company:

Approx. Seeds Left:

Reorder: ○ YES ○ NO

SEED INFORMATION

Seed Name:

Year on Packet:

Year Purchased:

Company:

Approx. Seeds Left:

Reorder: ○ YES ○ NO

SEED INFORMATION

Seed Name:

Year on Packet:

Year Purchased:

Company:

Approx. Seeds Left:

Reorder: ○ YES ○ NO

SEED INFORMATION

Seed Name:

Year on Packet:

Year Purchased:

Company:

Approx. Seeds Left:

Reorder: ○ YES ○ NO

SEED INFORMATION

Seed Name:

Year on Packet:

Year Purchased:

Company:

Approx. Seeds Left:

Reorder: ○ YES ○ NO

Seed Inventory

SEED INFORMATION

Seed Name:	
Year on Packet:	
Year Purchased:	
Company:	
Approx. Seeds Left:	
Reorder: ○ YES ○ NO	

SEED INFORMATION

Seed Name:	
Year on Packet:	
Year Purchased:	
Company:	
Approx. Seeds Left:	
Reorder: ○ YES ○ NO	

SEED INFORMATION

Seed Name:	
Year on Packet:	
Year Purchased:	
Company:	
Approx. Seeds Left:	
Reorder: ○ YES ○ NO	

SEED INFORMATION

Seed Name:	
Year on Packet:	
Year Purchased:	
Company:	
Approx. Seeds Left:	
Reorder: ○ YES ○ NO	

SEED INFORMATION

Seed Name:	
Year on Packet:	
Year Purchased:	
Company:	
Approx. Seeds Left:	
Reorder: ○ YES ○ NO	

SEED INFORMATION

Seed Name:	
Year on Packet:	
Year Purchased:	
Company:	
Approx. Seeds Left:	
Reorder: ○ YES ○ NO	

Seed Inventory

SEED INFORMATION

Seed Name:	
Year on Packet:	
Year Purchased:	
Company:	
Approx. Seeds Left:	
Reorder: ○ YES ○ NO	

SEED INFORMATION

Seed Name:	
Year on Packet:	
Year Purchased:	
Company:	
Approx. Seeds Left:	
Reorder: ○ YES ○ NO	

SEED INFORMATION

Seed Name:	
Year on Packet:	
Year Purchased:	
Company:	
Approx. Seeds Left:	
Reorder: ○ YES ○ NO	

SEED INFORMATION

Seed Name:	
Year on Packet:	
Year Purchased:	
Company:	
Approx. Seeds Left:	
Reorder: ○ YES ○ NO	

SEED INFORMATION

Seed Name:	
Year on Packet:	
Year Purchased:	
Company:	
Approx. Seeds Left:	
Reorder: ○ YES ○ NO	

SEED INFORMATION

Seed Name:	
Year on Packet:	
Year Purchased:	
Company:	
Approx. Seeds Left:	
Reorder: ○ YES ○ NO	

Seed Inventory

SEED INFORMATION

Seed Name:

Year on Packet:

Year Purchased:

Company:

Approx. Seeds Left:

Reorder: ○ YES ○ NO

SEED INFORMATION

Seed Name:

Year on Packet:

Year Purchased:

Company:

Approx. Seeds Left:

Reorder: ○ YES ○ NO

SEED INFORMATION

Seed Name:

Year on Packet:

Year Purchased:

Company:

Approx. Seeds Left:

Reorder: ○ YES ○ NO

SEED INFORMATION

Seed Name:

Year on Packet:

Year Purchased:

Company:

Approx. Seeds Left:

Reorder: ○ YES ○ NO

SEED INFORMATION

Seed Name:

Year on Packet:

Year Purchased:

Company:

Approx. Seeds Left:

Reorder: ○ YES ○ NO

SEED INFORMATION

Seed Name:

Year on Packet:

Year Purchased:

Company:

Approx. Seeds Left:

Reorder: ○ YES ○ NO

Seed Inventory

SEED INFORMATION

Seed Name:
Year on Packet:
Year Purchased:
Company:
Approx. Seeds Left:
Reorder: ○ YES ○ NO

SEED INFORMATION

Seed Name:
Year on Packet:
Year Purchased:
Company:
Approx. Seeds Left:
Reorder: ○ YES ○ NO

SEED INFORMATION

Seed Name:
Year on Packet:
Year Purchased:
Company:
Approx. Seeds Left:
Reorder: ○ YES ○ NO

SEED INFORMATION

Seed Name:
Year on Packet:
Year Purchased:
Company:
Approx. Seeds Left:
Reorder: ○ YES ○ NO

SEED INFORMATION

Seed Name:
Year on Packet:
Year Purchased:
Company:
Approx. Seeds Left:
Reorder: ○ YES ○ NO

SEED INFORMATION

Seed Name:
Year on Packet:
Year Purchased:
Company:
Approx. Seeds Left:
Reorder: ○ YES ○ NO

Seed Inventory

SEED INFORMATION

Seed Name:

Year on Packet:

Year Purchased:

Company:

Approx. Seeds Left:

Reorder: ○ YES ○ NO

SEED INFORMATION

Seed Name:

Year on Packet:

Year Purchased:

Company:

Approx. Seeds Left:

Reorder: ○ YES ○ NO

SEED INFORMATION

Seed Name:

Year on Packet:

Year Purchased:

Company:

Approx. Seeds Left:

Reorder: ○ YES ○ NO

SEED INFORMATION

Seed Name:

Year on Packet:

Year Purchased:

Company:

Approx. Seeds Left:

Reorder: ○ YES ○ NO

SEED INFORMATION

Seed Name:

Year on Packet:

Year Purchased:

Company:

Approx. Seeds Left:

Reorder: ○ YES ○ NO

SEED INFORMATION

Seed Name:

Year on Packet:

Year Purchased:

Company:

Approx. Seeds Left:

Reorder: ○ YES ○ NO

Seed Inventory

SEED INFORMATION

| Seed Name: |
| Year on Packet: |
| Year Purchased: |
| Company: |
| Approx. Seeds Left: |
| Reorder: ○ YES   ○ NO |

SEED INFORMATION

| Seed Name: |
| Year on Packet: |
| Year Purchased: |
| Company: |
| Approx. Seeds Left: |
| Reorder: ○ YES   ○ NO |

SEED INFORMATION

| Seed Name: |
| Year on Packet: |
| Year Purchased: |
| Company: |
| Approx. Seeds Left: |
| Reorder: ○ YES   ○ NO |

SEED INFORMATION

| Seed Name: |
| Year on Packet: |
| Year Purchased: |
| Company: |
| Approx. Seeds Left: |
| Reorder: ○ YES   ○ NO |

SEED INFORMATION

| Seed Name: |
| Year on Packet: |
| Year Purchased: |
| Company: |
| Approx. Seeds Left: |
| Reorder: ○ YES   ○ NO |

SEED INFORMATION

| Seed Name: |
| Year on Packet: |
| Year Purchased: |
| Company: |
| Approx. Seeds Left: |
| Reorder: ○ YES   ○ NO |

Seed Inventory

SEED INFORMATION

| Seed Name: |
| Year on Packet: |
| Year Purchased: |
| Company: |
| Approx. Seeds Left: |
| Reorder:　○ YES　○ NO |

SEED INFORMATION

| Seed Name: |
| Year on Packet: |
| Year Purchased: |
| Company: |
| Approx. Seeds Left: |
| Reorder:　○ YES　○ NO |

SEED INFORMATION

| Seed Name: |
| Year on Packet: |
| Year Purchased: |
| Company: |
| Approx. Seeds Left: |
| Reorder:　○ YES　○ NO |

SEED INFORMATION

| Seed Name: |
| Year on Packet: |
| Year Purchased: |
| Company: |
| Approx. Seeds Left: |
| Reorder:　○ YES　○ NO |

SEED INFORMATION

| Seed Name: |
| Year on Packet: |
| Year Purchased: |
| Company: |
| Approx. Seeds Left: |
| Reorder:　○ YES　○ NO |

SEED INFORMATION

| Seed Name: |
| Year on Packet: |
| Year Purchased: |
| Company: |
| Approx. Seeds Left: |
| Reorder:　○ YES　○ NO |

Seed Inventory

SEED INFORMATION

Seed Name:

Year on Packet:

Year Purchased:

Company:

Approx. Seeds Left:

Reorder: ○ YES ○ NO

SEED INFORMATION

Seed Name:

Year on Packet:

Year Purchased:

Company:

Approx. Seeds Left:

Reorder: ○ YES ○ NO

SEED INFORMATION

Seed Name:

Year on Packet:

Year Purchased:

Company:

Approx. Seeds Left:

Reorder: ○ YES ○ NO

SEED INFORMATION

Seed Name:

Year on Packet:

Year Purchased:

Company:

Approx. Seeds Left:

Reorder: ○ YES ○ NO

SEED INFORMATION

Seed Name:

Year on Packet:

Year Purchased:

Company:

Approx. Seeds Left:

Reorder: ○ YES ○ NO

SEED INFORMATION

Seed Name:

Year on Packet:

Year Purchased:

Company:

Approx. Seeds Left:

Reorder: ○ YES ○ NO

Seed Inventory

SEED INFORMATION

Seed Name:
Year on Packet:
Year Purchased:
Company:
Approx. Seeds Left:
Reorder: ○ YES   ○ NO

SEED INFORMATION

Seed Name:
Year on Packet:
Year Purchased:
Company:
Approx. Seeds Left:
Reorder: ○ YES   ○ NO

SEED INFORMATION

Seed Name:
Year on Packet:
Year Purchased:
Company:
Approx. Seeds Left:
Reorder: ○ YES   ○ NO

SEED INFORMATION

Seed Name:
Year on Packet:
Year Purchased:
Company:
Approx. Seeds Left:
Reorder: ○ YES   ○ NO

SEED INFORMATION

Seed Name:
Year on Packet:
Year Purchased:
Company:
Approx. Seeds Left:
Reorder: ○ YES   ○ NO

SEED INFORMATION

Seed Name:
Year on Packet:
Year Purchased:
Company:
Approx. Seeds Left:
Reorder: ○ YES   ○ NO

Seed Inventory

SEED INFORMATION

Seed Name:
Year on Packet:
Year Purchased:
Company:
Approx. Seeds Left:
Reorder: ○ YES ○ NO

SEED INFORMATION

Seed Name:
Year on Packet:
Year Purchased:
Company:
Approx. Seeds Left:
Reorder: ○ YES ○ NO

SEED INFORMATION

Seed Name:
Year on Packet:
Year Purchased:
Company:
Approx. Seeds Left:
Reorder: ○ YES ○ NO

SEED INFORMATION

Seed Name:
Year on Packet:
Year Purchased:
Company:
Approx. Seeds Left:
Reorder: ○ YES ○ NO

SEED INFORMATION

Seed Name:
Year on Packet:
Year Purchased:
Company:
Approx. Seeds Left:
Reorder: ○ YES ○ NO

SEED INFORMATION

Seed Name:
Year on Packet:
Year Purchased:
Company:
Approx. Seeds Left:
Reorder: ○ YES ○ NO

Seed Inventory

SEED INFORMATION

Seed Name:

Year on Packet:

Year Purchased:

Company:

Approx. Seeds Left:

Reorder: ○ YES ○ NO

SEED INFORMATION

Seed Name:

Year on Packet:

Year Purchased:

Company:

Approx. Seeds Left:

Reorder: ○ YES ○ NO

SEED INFORMATION

Seed Name:

Year on Packet:

Year Purchased:

Company:

Approx. Seeds Left:

Reorder: ○ YES ○ NO

SEED INFORMATION

Seed Name:

Year on Packet:

Year Purchased:

Company:

Approx. Seeds Left:

Reorder: ○ YES ○ NO

SEED INFORMATION

Seed Name:

Year on Packet:

Year Purchased:

Company:

Approx. Seeds Left:

Reorder: ○ YES ○ NO

SEED INFORMATION

Seed Name:

Year on Packet:

Year Purchased:

Company:

Approx. Seeds Left:

Reorder: ○ YES ○ NO

Seed Inventory

SEED INFORMATION

Seed Name:
Year on Packet:
Year Purchased:
Company:
Approx. Seeds Left:
Reorder: ○ YES   ○ NO

SEED INFORMATION

Seed Name:
Year on Packet:
Year Purchased:
Company:
Approx. Seeds Left:
Reorder: ○ YES   ○ NO

SEED INFORMATION

Seed Name:
Year on Packet:
Year Purchased:
Company:
Approx. Seeds Left:
Reorder: ○ YES   ○ NO

SEED INFORMATION

Seed Name:
Year on Packet:
Year Purchased:
Company:
Approx. Seeds Left:
Reorder: ○ YES   ○ NO

SEED INFORMATION

Seed Name:
Year on Packet:
Year Purchased:
Company:
Approx. Seeds Left:
Reorder: ○ YES   ○ NO

SEED INFORMATION

Seed Name:
Year on Packet:
Year Purchased:
Company:
Approx. Seeds Left:
Reorder: ○ YES   ○ NO

Seed Germination Log

SEED INFO

Name: _____
_____

Variety: _____
_____

PLANTING INFO

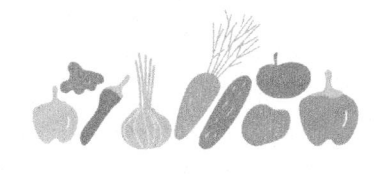

Plant Date: ____ / ____ / ____

Inside Planting: ☐   Direct Planting: ☐

Transplant Date: ____ / ____ / ____

% Germination (Approx): _____

BED INFO

Bed Name or #: _____

Plant/Seed Spacing: _____

Harvest Date: ____ / ____ / ____

Companion Plants:

_____ _____
_____ _____
_____ _____

Rotate With:
_____
_____

Plant Again?   YES ○   NO ○

SOIL USED

Speciality Soil: _____
_____

Amendments: _____
_____

Fertilizers: _____
_____

PESTS/PROBLEMS

_____
_____
_____
_____
_____
_____

NOTES

_____
_____
_____
_____
_____
_____

Seed Germination Log

SEED INFO

Name: _____
_____

Variety: _____
_____

PLANTING INFO

Plant Date: ____ / ____ / ____

Inside Planting: ☐   Direct Planting: ☐

Transplant Date: ____ / ____ / ____

% Germination (Approx): _____

BED INFO

Bed Name or #: _____

Plant/Seed Spacing: _____

Harvest Date: ____ / ____ / ____

Companion Plants:

_____ _____
_____ _____
_____ _____

Rotate With:
_____
_____

Plant Again?   YES ○   NO ○

SOIL USED

Speciality Soil: _____
_____

Amendments: _____
_____

Fertilizers: _____
_____

PESTS/PROBLEMS

_____
_____
_____
_____
_____

NOTES

_____
_____
_____
_____
_____
_____
_____

Seed Germination Log

SEED INFO

Name: _____
_____

Variety: _____
_____

PLANTING INFO

Plant Date: ____/____/____

Inside Planting: ☐   Direct Planting: ☐

Transplant Date: ____/____/____

% Germination (Approx): _____

BED INFO

Bed Name or #: _____

Plant/Seed Spacing: _____

Harvest Date: ____/____/____

Companion Plants:
_____ _____
_____ _____
_____ _____

Rotate With:
_____
_____

Plant Again?   YES ○   NO ○

SOIL USED

Speciality Soil: _____
_____

Amendments: _____
_____

Fertilizers: _____
_____

PESTS/PROBLEMS

_____
_____
_____
_____
_____
_____

NOTES

_____
_____
_____
_____
_____
_____

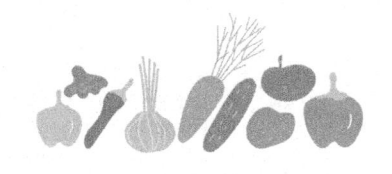

Seed Germination Log

SEED INFO

Name: _____
_____

Variety: _____
_____

PLANTING INFO

Plant Date: ___/___/___

Inside Planting: ☐ Direct Planting: ☐

Transplant Date: ___/___/___

% Germination (Approx): _____

BED INFO

Bed Name or #: _____

Plant/Seed Spacing: _____

Harvest Date: ___/___/___

Companion Plants:
_____ _____
_____ _____
_____ _____
_____ _____

Rotate With:
_____
_____

Plant Again?   YES ○   NO ○

SOIL USED

Speciality Soil: _____
_____

Amendments: _____
_____

Fertilizers: _____
_____

PESTS/PROBLEMS

_____
_____
_____
_____
_____
_____

NOTES

_____
_____
_____
_____
_____
_____

Seed Germination Log

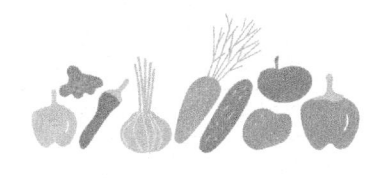

SEED INFO

Name: _____

_____

Variety: _____

_____

PLANTING INFO

Plant Date: ____ / ____ / ____

Inside Planting: ☐   Direct Planting: ☐

Transplant Date: ____ / ____ / ____

% Germination (Approx): _____

BED INFO

Bed Name or #: _____

Plant/Seed Spacing: _____

Harvest Date: ____ / ____ / ____

Companion Plants:

_____ _____
_____ _____
_____ _____

Rotate With:

_____
_____

Plant Again?   YES ○   NO ○

SOIL USED

Speciality Soil: _____
_____

Amendments: _____
_____

Fertilizers: _____
_____

PESTS/PROBLEMS

_____
_____
_____
_____
_____
_____

NOTES

_____
_____
_____
_____
_____
_____

Seed Germination Log

SEED INFO

Name: _____
_____

Variety: _____
_____

PLANTING INFO

Plant Date: ____/____/____

Inside Planting: ☐   Direct Planting: ☐

Transplant Date: ____/____/____

% Germination (Approx): _____

BED INFO

Bed Name or #: _____

Plant/Seed Spacing: _____

Harvest Date: ____/____/____

Companion Plants:

_____ _____
_____ _____
_____ _____

Rotate With:
_____
_____

Plant Again?   YES ○   NO ○

SOIL USED

Speciality Soil: _____
_____

Amendments: _____
_____

Fertilizers: _____
_____

PESTS/PROBLEMS

_____
_____
_____
_____
_____
_____

NOTES

_____
_____
_____
_____
_____
_____

Seed Germination Log

SEED INFO

Name: _____

Variety: _____

PLANTING INFO

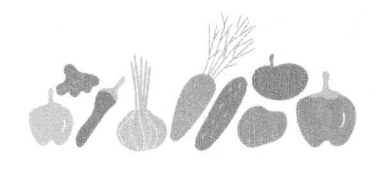

Plant Date: ____/____/____

Inside Planting: ☐   Direct Planting: ☐

Transplant Date: ____/____/____

% Germination (Approx): _____

BED INFO

Bed Name or #: _____

Plant/Seed Spacing: _____

Harvest Date: ____/____/____

Companion Plants:
_____ _____
_____ _____
_____ _____

Rotate With:
_____
_____

Plant Again?   YES ○   NO ○

SOIL USED

Speciality Soil: _____

Amendments: _____

Fertilizers: _____

PESTS/PROBLEMS

_____
_____
_____
_____
_____
_____

NOTES

_____
_____
_____
_____
_____
_____

Seed Germination Log

SEED INFO

Name: _____

_____

Variety: _____

_____

PLANTING INFO

Plant Date: ___/___/___

Inside Planting: ☐ Direct Planting: ☐

Transplant Date: ___/___/___

% Germination (Approx): _____

BED INFO

Bed Name or #: _____

Plant/Seed Spacing: _____

Harvest Date: ___/___/___

Companion Plants:

_____ _____

_____ _____

_____ _____

Rotate With:

_____

_____

Plant Again?   YES ○   NO ○

SOIL USED

Speciality Soil: _____

_____

Amendments: _____

_____

Fertilizers: _____

_____

PESTS/PROBLEMS

_____

_____

_____

_____

_____

NOTES

_____

_____

_____

_____

_____

_____

Seed Germination Log

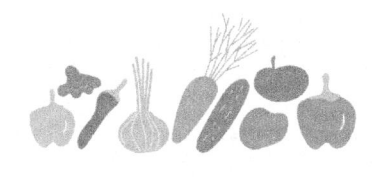

SEED INFO

Name: _____
_____

Variety: _____
_____

PLANTING INFO

Plant Date: ____/____/____

Inside Planting: ☐   Direct Planting: ☐

Transplant Date: ____/____/____

% Germination (Approx): _____

BED INFO

Bed Name or #: _____

Plant/Seed Spacing: _____

Harvest Date: ____/____/____

Companion Plants:
_____ _____
_____ _____
_____ _____

Rotate With:
_____
_____

Plant Again?   YES ○   NO ○

SOIL USED

Speciality Soil: _____
_____

Amendments: _____
_____

Fertilizers: _____
_____

PESTS/PROBLEMS

_____
_____
_____
_____
_____

NOTES

_____
_____
_____
_____
_____

Seed Germination Log

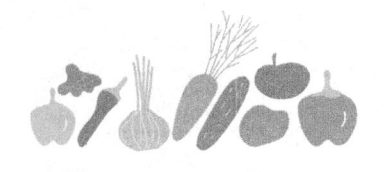

SEED INFO

Name: _____
_____

Variety: _____
_____

PLANTING INFO

Plant Date: ____ / ____ / ____

Inside Planting: ☐   Direct Planting: ☐

Transplant Date: ____ / ____ / ____

% Germination (Approx): _____

BED INFO

Bed Name or #: _____

Plant/Seed Spacing: _____

Harvest Date: ____ / ____ / ____

Companion Plants:

_____   _____
_____   _____
_____   _____

Rotate With:
_____
_____

Plant Again?   YES ○   NO ○

SOIL USED

Speciality Soil: _____
_____

Amendments: _____
_____

Fertilizers: _____
_____

PESTS/PROBLEMS

_____
_____
_____
_____
_____
_____

NOTES

_____
_____
_____
_____
_____
_____
_____

Seed Germination Log

SEED INFO

Name: _____

_____

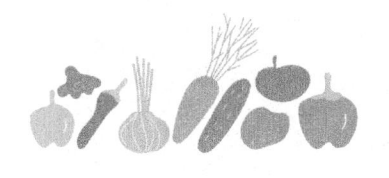

Variety: _____

_____

PLANTING INFO

Plant Date: ____/____/____

Inside Planting: ☐   Direct Planting: ☐

Transplant Date: ____/____/____

% Germination (Approx): _____

BED INFO

Bed Name or #: _____

Plant/Seed Spacing: _____

Harvest Date: ____/____/____

Companion Plants:

_____ _____
_____ _____
_____ _____

Rotate With:

_____
_____

Plant Again?   YES ○   NO ○

SOIL USED

Speciality Soil: _____
_____

Amendments: _____
_____

Fertilizers: _____
_____

PESTS/PROBLEMS

_____
_____
_____
_____
_____
_____

NOTES

_____
_____
_____
_____
_____
_____

Seed Germination Log

SEED INFO

Name: _____
_____

Variety: _____
_____

PLANTING INFO

Plant Date: ___/___/___

Inside Planting: ☐   Direct Planting: ☐

Transplant Date: ___/___/___

% Germination (Approx): _____

BED INFO

Bed Name or #: _____

Plant/Seed Spacing: _____

Harvest Date: ___/___/___

Companion Plants:
_____   _____
_____   _____
_____   _____

Rotate With:
_____
_____

Plant Again?   YES ○   NO ○

SOIL USED

Speciality Soil: _____
_____

Amendments: _____
_____

Fertilizers: _____
_____

PESTS/PROBLEMS

_____
_____
_____
_____
_____
_____

NOTES

_____
_____
_____
_____
_____
_____

31

Seed Germination Log

SEED INFO

Name: _____
_____

Variety: _____
_____

PLANTING INFO

Plant Date: ____/____/____

Inside Planting: ☐   Direct Planting: ☐

Transplant Date: ____/____/____

% Germination (Approx): _____

BED INFO

Bed Name or #: _____

Plant/Seed Spacing: _____

Harvest Date: ____/____/____

Companion Plants:

_____   _____
_____   _____
_____   _____

Rotate With:
_____
_____

Plant Again?   YES ○   NO ○

SOIL USED

Speciality Soil: _____
_____

Amendments: _____
_____

Fertilizers: _____
_____

PESTS/PROBLEMS

_____
_____
_____
_____
_____
_____

NOTES

_____
_____
_____
_____
_____
_____

Seed Germination Log

SEED INFO

Name: _____
_____

Variety: _____
_____

PLANTING INFO

Plant Date: ___ / ___ / ___

Inside Planting: ☐ Direct Planting: ☐

Transplant Date: ___ / ___ / ___

% Germination (Approx): _____

BED INFO

Bed Name or #: _____

Plant/Seed Spacing: _____

Harvest Date: ___ / ___ / ___

Companion Plants:

_____ _____
_____ _____
_____ _____

Rotate With:
_____
_____

Plant Again?   YES ○   NO ○

SOIL USED

Speciality Soil: _____
_____

Amendments: _____
_____

Fertilizers: _____
_____

PESTS/PROBLEMS

_____
_____
_____
_____
_____
_____

NOTES

_____
_____
_____
_____
_____
_____

Seed Germination Log

SEED INFO

Name: _____
_____

Variety: _____
_____

PLANTING INFO

Plant Date: ____ / ____ / ____

Inside Planting: ☐   Direct Planting: ☐

Transplant Date: ____ / ____ / ____

% Germination (Approx): _____

BED INFO

Bed Name or #: _____

Plant/Seed Spacing: _____

Harvest Date: ____ / ____ / ____

Companion Plants:

_____ _____
_____ _____
_____ _____

Rotate With:
_____
_____

Plant Again?   YES ○   NO ○

SOIL USED

Speciality Soil: _____
_____

Amendments: _____
_____

Fertilizers: _____
_____

PESTS/PROBLEMS

_____
_____
_____
_____
_____
_____

NOTES

_____
_____
_____
_____
_____
_____
_____

Seed Germination Log

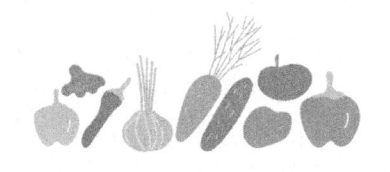

SEED INFO

Name: _____
_____

Variety: _____
_____

PLANTING INFO

Plant Date: ___ / ___ / ___

Inside Planting: ☐   Direct Planting: ☐

Transplant Date: ___ / ___ / ___

% Germination (Approx): _____

BED INFO

Bed Name or #: _____

Plant/Seed Spacing: _____

Harvest Date: ___ / ___ / ___

Companion Plants:

_____ _____
_____ _____
_____ _____

Rotate With:

_____
_____

Plant Again?   YES ○   NO ○

SOIL USED

Speciality Soil: _____
_____

Amendments: _____
_____

Fertilizers: _____
_____

PESTS/PROBLEMS

_____
_____
_____
_____
_____
_____

NOTES

_____
_____
_____
_____
_____
_____
_____

Seed Germination Log

SEED INFO

Name: _____

_____

Variety: _____

_____

PLANTING INFO

Plant Date: ____/____/____

Inside Planting: ☐   Direct Planting: ☐

Transplant Date: ____/____/____

% Germination (Approx): _____

BED INFO

Bed Name or #: _____

Plant/Seed Spacing: _____

Harvest Date: ____/____/____

Companion Plants:

_____ _____
_____ _____
_____ _____

Rotate With:

_____
_____

Plant Again?   YES ○   NO ○

SOIL USED

Speciality Soil: _____

_____

Amendments: _____

_____

Fertilizers: _____

_____

PESTS/PROBLEMS

_____
_____
_____
_____
_____
_____

NOTES

_____
_____
_____
_____
_____
_____

Seed Germination Log

SEED INFO

Name: _____
_____

Variety: _____
_____

PLANTING INFO

Plant Date: _____ / _____ / _____

Inside Planting: ☐   Direct Planting: ☐

Transplant Date: _____ / _____ / _____

% Germination (Approx): _____

BED INFO

Bed Name or #: _____

Plant/Seed Spacing: _____

Harvest Date: _____ / _____ / _____

Companion Plants:

_____ _____
_____ _____
_____ _____

Rotate With:

_____
_____

Plant Again?   YES ○   NO ○

SOIL USED

Speciality Soil: _____
_____

Amendments: _____
_____

Fertilizers: _____
_____

PESTS/PROBLEMS

_____
_____
_____
_____
_____
_____

NOTES

_____
_____
_____
_____
_____
_____
_____

Seed Germination Log

SEED INFO

Name: _____
_____

Variety: _____
_____

PLANTING INFO

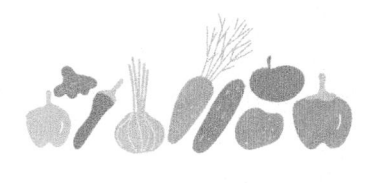

Plant Date: ____/____/____

Inside Planting: ☐   Direct Planting: ☐

Transplant Date: ____/____/____

% Germination (Approx): _____

BED INFO

Bed Name or #: _____

Plant/Seed Spacing: _____

Harvest Date: ____/____/____

Companion Plants:

_____ _____
_____ _____
_____ _____

Rotate With:
_____
_____

Plant Again?   YES ○   NO ○

SOIL USED

Speciality Soil: _____
_____

Amendments: _____
_____

Fertilizers: _____
_____

PESTS/PROBLEMS

_____
_____
_____
_____
_____
_____

NOTES

_____
_____
_____
_____
_____
_____

Seed Germination Log

SEED INFO

Name: _____
_____

Variety: _____
_____

PLANTING INFO

Plant Date: ___/___/___

Inside Planting: ☐   Direct Planting: ☐

Transplant Date: ___/___/___

% Germination (Approx): _____

BED INFO

Bed Name or #: _____

Plant/Seed Spacing: _____

Harvest Date: ___/___/___

Companion Plants:

_____ _____
_____ _____
_____ _____

Rotate With:

_____
_____

Plant Again?   YES ○   NO ○

SOIL USED

Speciality Soil: _____
_____

Amendments: _____
_____

Fertilizers: _____
_____

PESTS/PROBLEMS

_____
_____
_____
_____
_____
_____

NOTES

_____
_____
_____
_____
_____
_____
_____

Seed Germination Log

SEED INFO

Name: _____
_____

Variety: _____
_____

PLANTING INFO

Plant Date: ____ / ____ / ____

Inside Planting: ☐   Direct Planting: ☐

Transplant Date: ____ / ____ / ____

% Germination (Approx): _____

BED INFO

Bed Name or #: _____

Plant/Seed Spacing: _____

Harvest Date: ____ / ____ / ____

Companion Plants:
_____ _____
_____ _____
_____ _____

Rotate With:
_____
_____

Plant Again?   YES ○   NO ○

SOIL USED

Speciality Soil: _____
_____

Amendments: _____
_____

Fertilizers: _____
_____

PESTS/PROBLEMS

_____
_____
_____
_____
_____
_____

NOTES

_____
_____
_____
_____
_____
_____

Seed Germination Log

SEED INFO

Name: _____
_____

Variety: _____
_____

PLANTING INFO

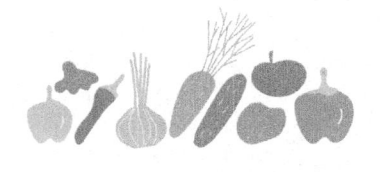

Plant Date: ____/____/____

Inside Planting: ☐   Direct Planting: ☐

Transplant Date: ____/____/____

% Germination (Approx): _____

BED INFO

Bed Name or #: _____

Plant/Seed Spacing: _____

Harvest Date: ____/____/____

Companion Plants:

_____   _____
_____   _____
_____   _____

Rotate With:
_____
_____

Plant Again?   YES ○   NO ○

SOIL USED

Speciality Soil: _____
_____

Amendments: _____
_____

Fertilizers: _____
_____

PESTS/PROBLEMS

_____
_____
_____
_____
_____
_____

NOTES

_____
_____
_____
_____
_____
_____

Seed Germination Log

SEED INFO

Name: _____
_____

Variety: _____
_____

PLANTING INFO

Plant Date: ____/____/____

Inside Planting: ☐   Direct Planting: ☐

Transplant Date: ____/____/____

% Germination (Approx): _____

BED INFO

Bed Name or #: _____

Plant/Seed Spacing: _____

Harvest Date: ____/____/____

Companion Plants:
_____ _____
_____ _____
_____ _____

Rotate With:
_____
_____

Plant Again?   YES ○   NO ○

SOIL USED

Speciality Soil: _____
_____

Amendments: _____
_____

Fertilizers: _____
_____

PESTS/PROBLEMS

_____
_____
_____
_____
_____
_____

NOTES

_____
_____
_____
_____
_____
_____
_____

Seed Germination Log

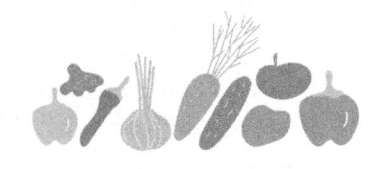

SEED INFO

Name: _____
_____

Variety: _____
_____

PLANTING INFO

Plant Date: ____ / ____ / ____

Inside Planting: ☐   Direct Planting: ☐

Transplant Date: ____ / ____ / ____

% Germination (Approx): _____

BED INFO

Bed Name or #: _____

Plant/Seed Spacing: _____

Harvest Date: ____ / ____ / ____

Companion Plants:

_____ _____
_____ _____
_____ _____

Rotate With:

_____
_____

Plant Again?   YES ○   NO ○

SOIL USED

Speciality Soil: _____
_____

Amendments: _____
_____

Fertilizers: _____
_____

PESTS/PROBLEMS

_____
_____
_____
_____
_____
_____

NOTES

_____
_____
_____
_____
_____
_____

Seed Germination Log

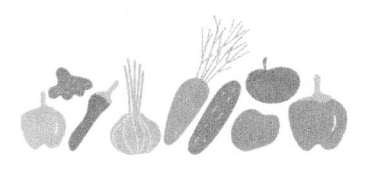

SEED INFO

Name: _____
_____

Variety: _____
_____

PLANTING INFO

Plant Date: _____ / _____ / _____

Inside Planting: ☐   Direct Planting: ☐

Transplant Date: _____ / _____ / _____

% Germination (Approx): _____

BED INFO

Bed Name or #: _____

Plant/Seed Spacing: _____

Harvest Date: _____ / _____ / _____

Companion Plants:

_____   _____
_____   _____
_____   _____
_____   _____

Rotate With:
_____
_____

Plant Again?   YES ○   NO ○

SOIL USED

Speciality Soil: _____
_____

Amendments: _____
_____

Fertilizers: _____
_____

PESTS/PROBLEMS

_____
_____
_____
_____
_____
_____

NOTES

_____
_____
_____
_____
_____
_____
_____

Seed Germination Log

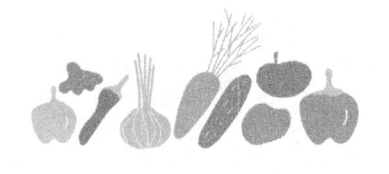

SEED INFO

Name: _____
_____

Variety: _____
_____

PLANTING INFO

Plant Date: ____/____/____

Inside Planting: ☐   Direct Planting: ☐

Transplant Date: ____/____/____

% Germination (Approx): _____

BED INFO

Bed Name or #: _____

Plant/Seed Spacing: _____

Harvest Date: ____/____/____

Companion Plants:
_____ _____
_____ _____
_____ _____

Rotate With:
_____
_____

Plant Again?   YES ○   NO ○

SOIL USED

Speciality Soil: _____
_____

Amendments: _____
_____

Fertilizers: _____
_____

PESTS/PROBLEMS

_____
_____
_____
_____
_____
_____

NOTES

_____
_____
_____
_____
_____
_____

Seed Germination Log

SEED INFO

Name: _____
_____

Variety: _____
_____

PLANTING INFO

Plant Date: ____ / ____ / ____

Inside Planting: ☐   Direct Planting: ☐

Transplant Date: ____ / ____ / ____

% Germination (Approx): _____

BED INFO

Bed Name or #: _____

Plant/Seed Spacing: _____

Harvest Date: ____ / ____ / ____

Companion Plants:
_____   _____
_____   _____
_____   _____

Rotate With:
_____
_____

Plant Again?   YES ○   NO ○

SOIL USED

Speciality Soil: _____
_____

Amendments: _____
_____

Fertilizers: _____
_____

PESTS/PROBLEMS

_____
_____
_____
_____
_____
_____

NOTES

_____
_____
_____
_____
_____
_____

Seed Germination Log

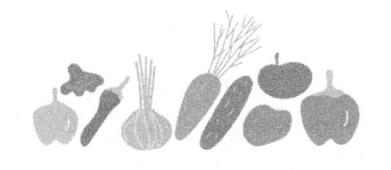

SEED INFO

Name: _____
_____

Variety: _____
_____

PLANTING INFO

Plant Date: ___/___/___

Inside Planting: ☐   Direct Planting: ☐

Transplant Date: ___/___/___

% Germination (Approx): _____

BED INFO

Bed Name or #: _____

Plant/Seed Spacing: _____

Harvest Date: ___/___/___

Companion Plants:
_____   _____
_____   _____
_____   _____

Rotate With:
_____
_____

Plant Again?   YES ◯   NO ◯

SOIL USED

Speciality Soil: _____
_____

Amendments: _____
_____

Fertilizers: _____
_____

PESTS/PROBLEMS

_____
_____
_____
_____
_____
_____

NOTES

_____
_____
_____
_____
_____
_____

Seed Germination Log

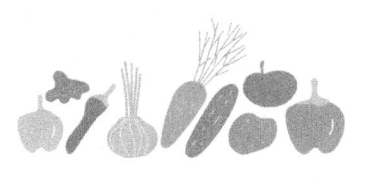

SEED INFO

Name: _____
_____

Variety: _____
_____

PLANTING INFO

Plant Date: ____/____/____

Inside Planting: ☐   Direct Planting: ☐

Transplant Date: ____/____/____

% Germination (Approx): _____

BED INFO

Bed Name or #: _____

Plant/Seed Spacing: _____

Harvest Date: ____/____/____

Companion Plants:

_____   _____
_____   _____
_____   _____

Rotate With:
_____
_____

Plant Again?   YES ○   NO ○

SOIL USED

Speciality Soil: _____
_____

Amendments: _____
_____

Fertilizers: _____
_____

PESTS/PROBLEMS

_____
_____
_____
_____
_____
_____

NOTES

_____
_____
_____
_____
_____
_____

48

Seed Germination Log

SEED INFO

Name: _____
_____

Variety: _____
_____

PLANTING INFO

Plant Date: _____ / _____ / _____

Inside Planting: ☐   Direct Planting: ☐

Transplant Date: _____ / _____ / _____

% Germination (Approx): _____

BED INFO

Bed Name or #: _____

Plant/Seed Spacing: _____

Harvest Date: _____ / _____ / _____

Companion Plants:

_____ _____
_____ _____
_____ _____

Rotate With:
_____
_____

Plant Again?   YES ○   NO ○

SOIL USED

Speciality Soil: _____
_____

Amendments: _____
_____

Fertilizers: _____
_____

PESTS/PROBLEMS

_____
_____
_____
_____
_____
_____
_____

NOTES

_____
_____
_____
_____
_____
_____

Garden Bed
INFORMATION

Bed Name/#: _____

SEASON	CROP	DATE PLANTED	SEEDS OR STARTS	ROTATED CROP	NOTES

Garden Bed
INFORMATION

Bed Name/#: _____

SEASON	CROP	DATE PLANTED	SEEDS OR STARTS	ROTATED CROP	NOTES

Garden Bed
INFORMATION

Bed Name/#: _____

SEASON	CROP	DATE PLANTED	SEEDS OR STARTS	ROTATED CROP	NOTES

Garden Bed
INFORMATION

Bed Name/#: _____

SEASON	CROP	DATE PLANTED	SEEDS OR STARTS	ROTATED CROP	NOTES

Garden Bed
INFORMATION

Bed Name/#: _____

SEASON	CROP	DATE PLANTED	SEEDS OR STARTS	ROTATED CROP	NOTES

Garden Bed
INFORMATION

Bed Name/#: _____

SEASON	CROP	DATE PLANTED	SEEDS OR STARTS	ROTATED CROP	NOTES

Garden Bed
INFORMATION

Bed Name/#: _____

SEASON	CROP	DATE PLANTED	SEEDS OR STARTS	ROTATED CROP	NOTES

Garden Bed
INFORMATION

Bed Name/#: _____

SEASON	CROP	DATE PLANTED	SEEDS OR STARTS	ROTATED CROP	NOTES

Garden Bed
INFORMATION

Bed Name/#: _____

SEASON	CROP	DATE PLANTED	SEEDS OR STARTS	ROTATED CROP	NOTES

Garden Bed
INFORMATION

Bed Name/#: _____

SEASON	CROP	DATE PLANTED	SEEDS OR STARTS	ROTATED CROP	NOTES

Garden Bed
INFORMATION

Bed Name/#: _____

SEASON	CROP	DATE PLANTED	SEEDS OR STARTS	ROTATED CROP	NOTES

Garden Bed
INFORMATION

Bed Name/#: _____

SEASON	CROP	DATE PLANTED	SEEDS OR STARTS	ROTATED CROP	NOTES

Garden Bed
INFORMATION

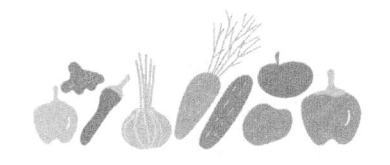

Bed Name/#: _____

SEASON	CROP	DATE PLANTED	SEEDS OR STARTS	ROTATED CROP	NOTES

Garden Bed
INFORMATION

Bed Name/#: _____

SEASON	CROP	DATE PLANTED	SEEDS OR STARTS	ROTATED CROP	NOTES

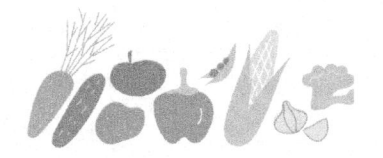

Garden Bed
INFORMATION

Bed Name/#: _____

SEASON	CROP	DATE PLANTED	SEEDS OR STARTS	ROTATED CROP	NOTES

Garden Bed
INFORMATION

Bed Name/#: _____

SEASON	CROP	DATE PLANTED	SEEDS OR STARTS	ROTATED CROP	NOTES

Garden Bed Information

Bed Name/#: _____

SEASON	CROP	DATE PLANTED	SEEDS OR STARTS	ROTATED CROP	NOTES

Garden Bed
INFORMATION

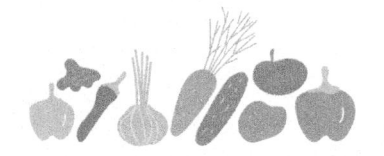

Bed Name/#: _____

SEASON	CROP	DATE PLANTED	SEEDS OR STARTS	ROTATED CROP	NOTES

Garden Bed
INFORMATION

Bed Name/#: _____

SEASON	CROP	DATE PLANTED	SEEDS OR STARTS	ROTATED CROP	NOTES

Garden Bed
INFORMATION

Bed Name/#: _____

SEASON	CROP	DATE PLANTED	SEEDS OR STARTS	ROTATED CROP	NOTES

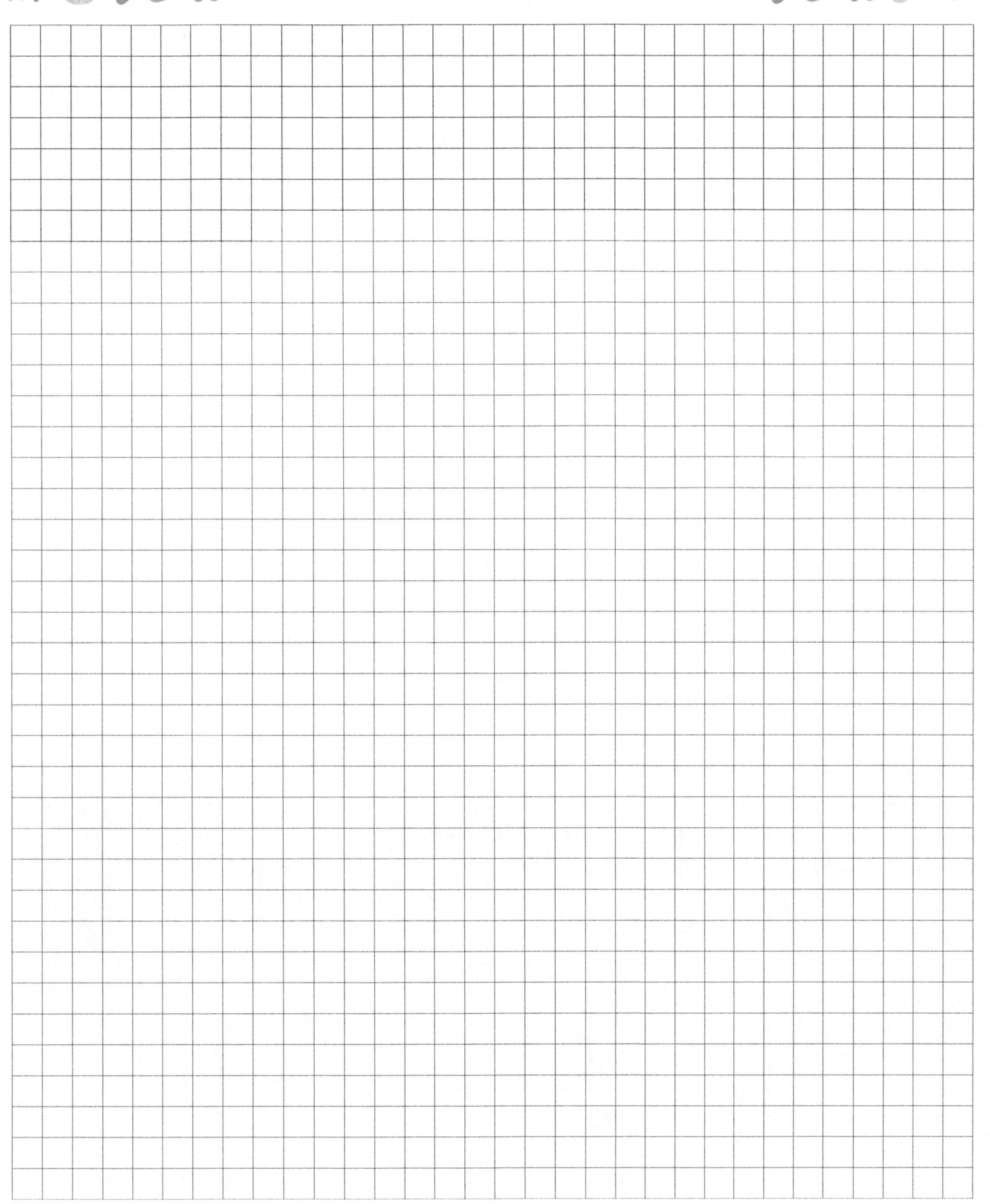

Garden Design Layout

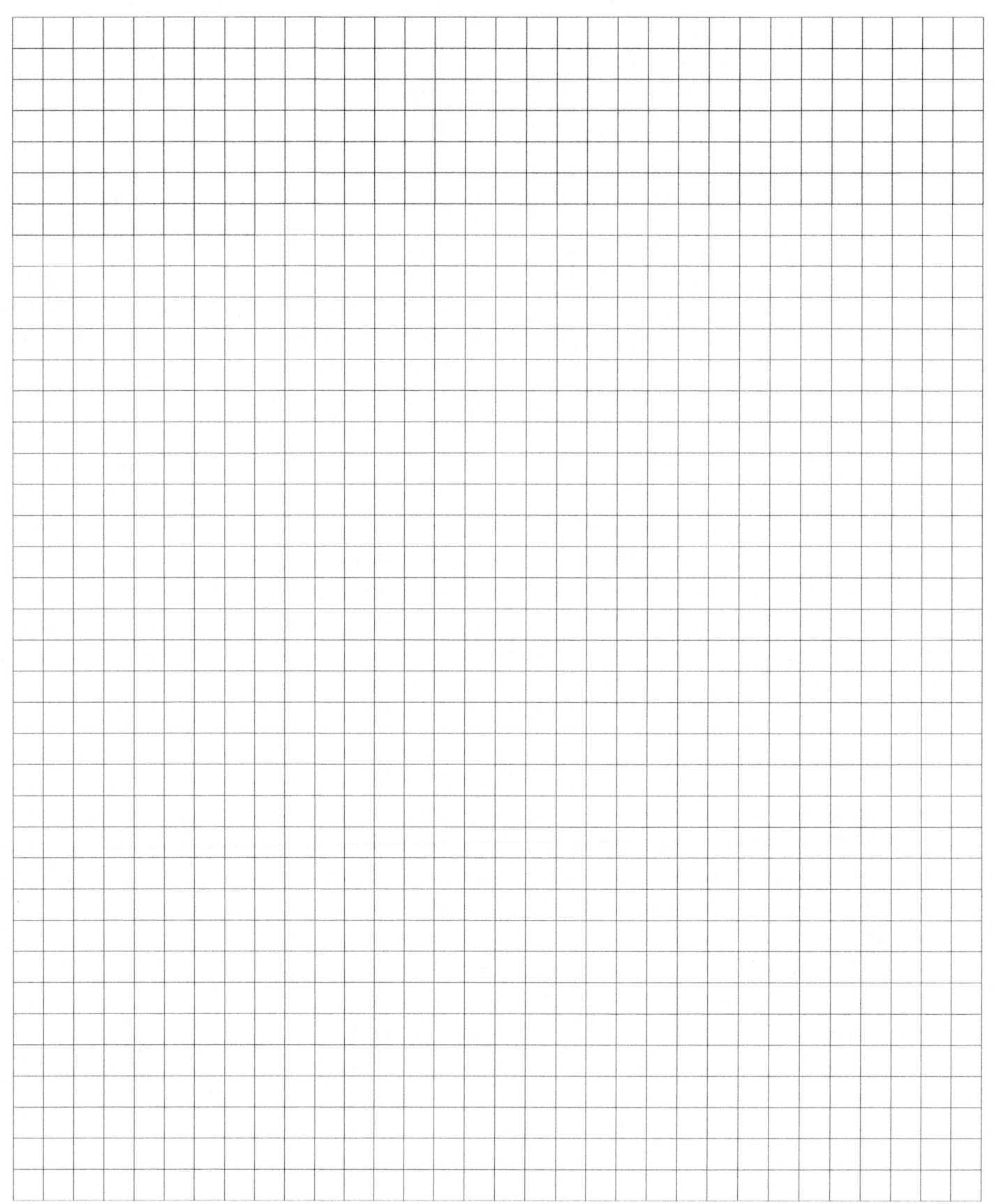

Garden Design Layout

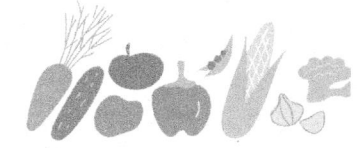

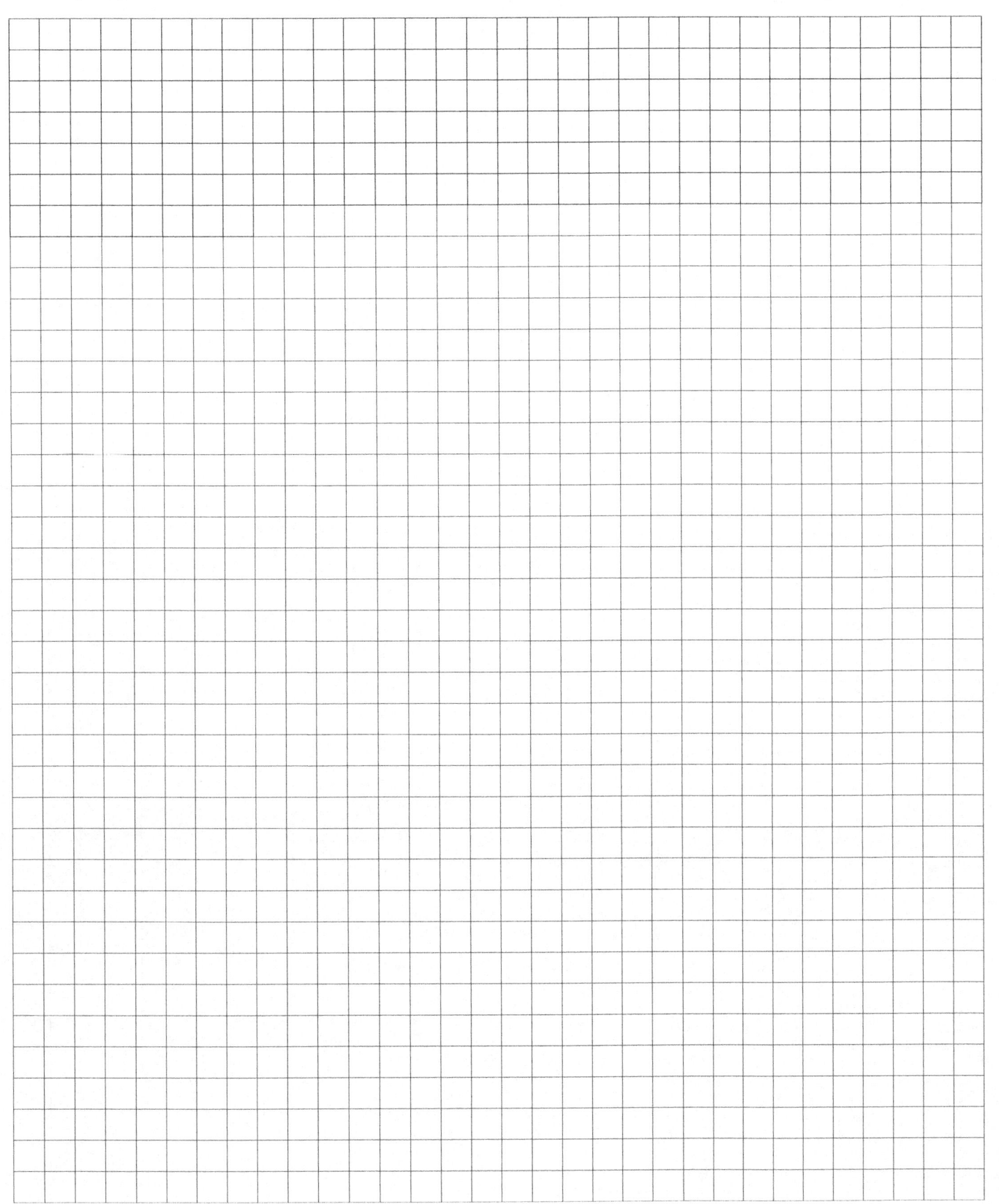

Garden Design
LAYOUT

Garden Design
LAYOUT

Garden Design
SQUARE FOOT

Garden Design
SQUARE FOOT

Garden Design
SQUARE FOOT

Garden Design
SQUARE FOOT

Flower Garden
LAYOUT

Flower Garden
LAYOUT

Flower Garden
LAYOUT

Flower Garden
LAYOUT

Flower Garden
SQUARE FOOT

Flower Garden
SQUARE FOOT

Flower Garden
SQUARE FOOT

Flower Garden
SQUARE FOOT

Planting Calendar

YEAR: _____

JANUARY

FEBRUARY

MARCH

APRIL

MAY

JUNE

JULY

AUGUST

SEPTEMBER

OCTOBER

NOVEMBER

DECEMBER

Planting Calendar

YEAR: _____

| JANUARY | FEBRUARY | MARCH |

| APRIL | MAY | JUNE |

| JULY | AUGUST | SEPTEMBER |

| OCTOBER | NOVEMBER | DECEMBER |

Seasonal Checklist

SPRING

☐ _____
☐ _____
☐ _____
☐ _____
☐ _____
☐ _____
☐ _____
☐ _____
☐ _____
☐ _____
☐ _____

SUMMER

☐ _____
☐ _____
☐ _____
☐ _____
☐ _____
☐ _____
☐ _____
☐ _____
☐ _____
☐ _____
☐ _____

FALL

☐ _____
☐ _____
☐ _____
☐ _____
☐ _____
☐ _____
☐ _____
☐ _____
☐ _____
☐ _____
☐ _____

WINTER

☐ _____
☐ _____
☐ _____
☐ _____
☐ _____
☐ _____
☐ _____
☐ _____
☐ _____
☐ _____
☐ _____

Seasonal Checklist

SPRING

- [ ]
- [ ]
- [ ]
- [ ]
- [ ]
- [ ]
- [ ]
- [ ]
- [ ]
- [ ]

SUMMER

- [ ]
- [ ]
- [ ]
- [ ]
- [ ]
- [ ]
- [ ]
- [ ]
- [ ]
- [ ]

FALL

- [ ]
- [ ]
- [ ]
- [ ]
- [ ]
- [ ]
- [ ]
- [ ]
- [ ]
- [ ]

WINTER

- [ ]
- [ ]
- [ ]
- [ ]
- [ ]
- [ ]
- [ ]
- [ ]
- [ ]
- [ ]

Garden To-Do List

- [ ] _____
- [ ] _____
- [ ] _____
- [ ] _____
- [ ] _____
- [ ] _____
- [ ] _____
- [ ] _____
- [ ] _____
- [ ] _____
- [ ] _____
- [ ] _____
- [ ] _____
- [ ] _____

Garden To-Do List

- [ ] _____
- [ ] _____
- [ ] _____
- [ ] _____
- [ ] _____
- [ ] _____
- [ ] _____
- [ ] _____
- [ ] _____
- [ ] _____
- [ ] _____
- [ ] _____
- [ ] _____
- [ ] _____
- [ ] _____

Garden To-Do List

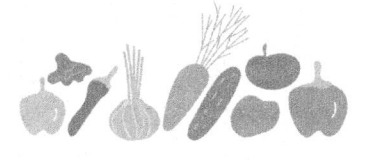

Garden To-Do List

Garden To-Do List

- [ ] _____
- [ ] _____
- [ ] _____
- [ ] _____
- [ ] _____
- [ ] _____
- [ ] _____
- [ ] _____
- [ ] _____
- [ ] _____
- [ ] _____
- [ ] _____
- [ ] _____
- [ ] _____
- [ ] _____

Garden To-Do List

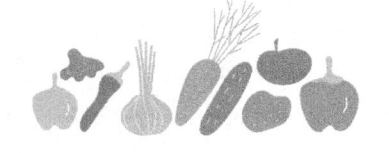

- [ ] _____
- [ ] _____
- [ ] _____
- [ ] _____
- [ ] _____
- [ ] _____
- [ ] _____
- [ ] _____
- [ ] _____
- [ ] _____
- [ ] _____
- [ ] _____
- [ ] _____
- [ ] _____
- [ ] _____

Garden To-Do List

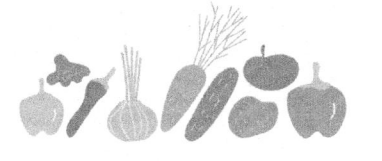

Garden To-Do List

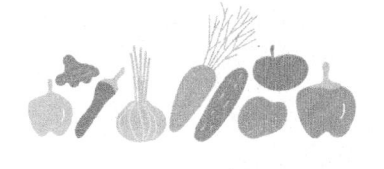

Supplier Information

Store Name:

Contact: _____

Phone: _____

Address: _____

_____

Products: _____

_____

E-Mail: _____

Website: _____

Rating: ☆ ☆ ☆ ☆ ☆

Store Name:

Contact: _____

Phone: _____

Address: _____

_____

Products: _____

_____

E-Mail: _____

Website: _____

Rating: ☆ ☆ ☆ ☆ ☆

Store Name:

Contact: _____

Phone: _____

Address: _____

_____

Products: _____

_____

E-Mail: _____

Website: _____

Rating: ☆ ☆ ☆ ☆ ☆

Supplier Information

Store Name: _____

Contact: _____　　Products: _____

Phone: _____　　　　_____

Address: _____　　E-Mail: _____

_____　　Website: _____

_____　　Rating: ☆ ☆ ☆ ☆ ☆

Store Name: _____

Contact: _____　　Products: _____

Phone: _____　　　　_____

Address: _____　　E-Mail: _____

_____　　Website: _____

_____　　Rating: ☆ ☆ ☆ ☆ ☆

Store Name: _____

Contact: _____　　Products: _____

Phone: _____　　　　_____

Address: _____　　E-Mail: _____

_____　　Website: _____

_____　　Rating: ☆ ☆ ☆ ☆ ☆

Supplier Information

Store Name:

Contact: _____

Phone: _____

Address: _____

_____

_____

Products: _____

_____

E-Mail: _____

Website: _____

Rating: ☆ ☆ ☆ ☆ ☆

Store Name:

Contact: _____

Phone: _____

Address: _____

_____

_____

Products: _____

_____

E-Mail: _____

Website: _____

Rating: ☆ ☆ ☆ ☆ ☆

Store Name:

Contact: _____

Phone: _____

Address: _____

_____

_____

Products: _____

_____

E-Mail: _____

Website: _____

Rating: ☆ ☆ ☆ ☆ ☆

Supplier Information

Store Name: _____

Contact: _____
Phone: _____
Address: _____
_____
_____

Products: _____
_____
E-Mail: _____
Website: _____
Rating: ☆ ☆ ☆ ☆ ☆

Store Name: _____

Contact: _____
Phone: _____
Address: _____
_____
_____

Products: _____
_____
E-Mail: _____
Website: _____
Rating: ☆ ☆ ☆ ☆ ☆

Store Name: _____

Contact: _____
Phone: _____
Address: _____
_____
_____

Products: _____
_____
E-Mail: _____
Website: _____
Rating: ☆ ☆ ☆ ☆ ☆

Supplier Information

Store Name:

Contact: _____

Phone: _____

Address: _____

_____

_____

Products: _____

_____

E-Mail: _____

Website: _____

Rating: ☆ ☆ ☆ ☆ ☆

Store Name:

Contact: _____

Phone: _____

Address: _____

_____

_____

Products: _____

_____

E-Mail: _____

Website: _____

Rating: ☆ ☆ ☆ ☆ ☆

Store Name:

Contact: _____

Phone: _____

Address: _____

_____

_____

Products: _____

_____

E-Mail: _____

Website: _____

Rating: ☆ ☆ ☆ ☆ ☆

Supplier Information

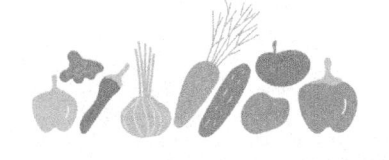

Store Name: _____

Contact: _____ Products: _____

Phone: _____ _____

Address: _____ E-Mail: _____

_____ Website: _____

_____ Rating: ☆ ☆ ☆ ☆ ☆

Store Name: _____

Contact: _____ Products: _____

Phone: _____ _____

Address: _____ E-Mail: _____

_____ Website: _____

_____ Rating: ☆ ☆ ☆ ☆ ☆

Store Name: _____

Contact: _____ Products: _____

Phone: _____ _____

Address: _____ E-Mail: _____

_____ Website: _____

_____ Rating: ☆ ☆ ☆ ☆ ☆

Garden Notes

DATE: _____ / _____ / _____

Garden Notes

DATE: ____/____/____

Garden Notes

DATE: ____/____/____

Garden Notes

DATE: ____ / ____ / ____

Garden Notes

DATE: ____ / ____ / ____

Garden Notes

DATE: ___ / ___ / ___

Garden Notes

DATE: _____/_____/_____

Garden Notes

DATE: ____ / ____ / ____

Garden Notes

DATE: ____ / ____ / ____

Garden Notes

DATE: _____ / _____ / _____

Thank you for purchasing

GARDEN PLANNER
and
SEED INVENTORY JOURNAL

If you enjoyed this journal please leave us a review on **Amazon** as we love to hear your feedback!

Find more great journals, activity books and coloring books at:

www.hometreespress.com

HOMETREES PRESS
Ojai • California

Made in the USA
Coppell, TX
25 June 2024